Computing and Systems Technology Division 2022

Held at the 2022 AIChE Spring Meeting and 18th Global Congress on Process Safety

San Antonio, Texas, USA
10 – 14 April 2022

ISBN: 978-1-7138-6828-6

Printed from e-media with permission by:

Curran Associates, Inc.
57 Morehouse Lane
Red Hook, NY 12571

Some format issues inherent in the e-media version may also appear in this print version.

Printed with permission by Curran Associates, Inc. (2023)

For permission requests, please contact AIChE
at the address below.

AIChE
120 Wall Street, FL 23
New York, NY 10005-4020

Phone: (800) 242-4363
Fax: (203) 775-5177

www.aiche.org

Additional copies of this publication are available from:

Curran Associates, Inc.
57 Morehouse Lane
Red Hook, NY 12571 USA
Phone: 845-758-0400
Fax: 845-758-2633
Email: curran@proceedings.com
Web: www.proceedings.com

TABLE OF CONTENTS

639464 Systematic Incorporation of Safety Assessment in Process Design, Intensification, and Control

Abstract:

The recent initiatives towards modular chemical process intensification (MCPI) have posed new challenges and opportunities to process safety evaluation and management [1-2]. The increasing plant complexity and interactivity with novel modular and task-integrated unit operations necessitate advances to accurately quantify the impact of new equipment designs on plant safety and to effectively control the safety performance during real-time operation [3-4]. In view of this, several systems-based methods have been proposed to incorporate inherent safety index as a primary design objective or to apply model predictive control for operational safety [5-8]. Despite these efforts, key research gaps remain on the lack of: (i) a fundamental understanding of intensification and modularization impacts on process safety, and (ii) a systematic approach to generate optimal MCPI design and operational strategies with guaranteed safety performance under disturbances.

To address these challenges, in this work we will first investigate a series of comparative case studies to rigorously analyze the process safety performance in modular and intensified designs. Representative safety metrics [9-12] are tested to quantify process safety at the following steady-state and dynamic operation scenarios: (i) evaluation of different methyl tertiary butyl ether (MTBE) reactive distillation designs against inherent safety principles, (ii) design optimization of a benzene nitration reaction system with safety and modularization considerations, and (iii) dynamic safety performance of an intensified reactive distillation process vs. a conventional reactor-distillation-recycle process with model predictive control. Based on this, we further propose a safety-aware explicit/multi-parametric model predictive control (mp-MPC) strategy [13-14]. The SWeHI index value [11] is selected according to the MCPI analyses and integrated in the parametric space to jointly determine the optimal control actions and to operate the process at a desired level of safety. Additional safety requirements, e.g. on process temperature, can be explicitly formulated as mp-MPC path constraints to identify the maximum set of disturbances and optimal set point selection to theoretically prevent any constraint violation through operation. The extension of this approach for safety-oriented simultaneous design and control optimization will also be discussed based on our recent work in intensified process systems [15]. The proposed approach will be demonstrated on a continuous stirred tank reactor case study for the processing of methylcyclopentadienyl manganese tricarbonyl at T2 Laboratories.

References

[1] Etchells, J. C. (2005). Process intensification: safety pros and cons. Process Safety and Environmental Protection, 83(2), 85-89.

[2] Yelvington, P., Gokhale, A., Grieco, W., Nara, L. (2019) The link between process safety and process intensification. Available from: https://www.aiche.org/academy/webinars/link-between-process-safety-and-process-intensification.

[3] Park, S., Xu, S., Rogers, W., Pasman, H., & El-Halwagi, M. M. (2020). Incorporating inherent safety during the conceptual process design stage: A literature review. Journal of Loss Prevention in the Process Industries, 63, 104040.

[4] Pistikopoulos, E. N., Tian, Y., & Bindlish, R. (2021). Operability and control in process intensification and modular design: Challenges and opportunities. AIChE Journal, 67(5), e17204.

[5] Castillo-Landero, A., Ortiz-Espinoza, A. P., & Jiménez-Gutiérrez, A. (2018). A process intensification methodology including economic, sustainability, and safety considerations. Industrial & Engineering Chemistry Research, 58(15), 6080-6092.

[6] Leveson, N. G., & Stephanopoulos, G. (2014). A system-theoretic, control-inspired view and approach to process safety. AIChE Journal, 60(1), 2-14.

[7] Albalawi, F., Durand, H., Alanqar, A., & Christofides, P. D. (2018). Achieving operational process safety via model predictive control. Journal of Loss Prevention in the Process Industries, 53, 74-88.

[8] Venkidasalapathy, J. A., & Kravaris, C. (2020). Safety-centered process control design based on dynamic safe set. Journal of Loss Prevention in the Process Industries, 65, 104126.

[9] American Institute of Chemical Engineers. (2010). Dow's Fire & Explosion Index Hazard Classification Guide, Wiley.

[10] Marshall, J. T. & Mundt, A. (1995). Dow's chemical exposure index guide. Process Safety Progress, 14, 163-170.

[11] Khan, F. I., Husain, T. & Abbasi, S. A. (2001). Safety weighted hazard index (SWeHI): A new, user-friendly tool for swift yet comprehensive hazard identification and safety evaluation in chemical process industry. Process Safety and Environmental Protection, 79, 65-80.

[12] Stoffen, P. G. (2005). Guidelines for quantitative risk assessment, Ministerievan Volkshuisvesting Ruimtelijke Ordening en Milieu. CPR E 18.

[13] Pistikopoulos, E. N., Diangelakis, N. A., & Oberdieck, R. (2020). Multi-parametric Optimization and Control. John Wiley & Sons.

[14] Pistikopoulos, E. N., Diangelakis, N. A., Oberdieck, R., Papathanasiou, M. M., Nascu, I., & Sun, M. (2015). PAROC – An integrated framework and software platform for the optimisation and advanced model-based control of process systems. Chemical Engineering Science, 136, 115-138.

[15] Tian, Y., Pappas, I., Burnak, B., Katz, J., & Pistikopoulos, E. N. (2021). Simultaneous design & control of a reactive distillation system – A parametric optimization & control approach. Chemical Engineering Science, 230, 116232.

639986 Analysis of Lithium-Ion Battery Management System and Degradation Mechanism

Abstract:

Battery technologies, particularly lithium-ion batteries (LIB) have shown promise in automotive applications. To ensure reliable, and safe operation, a battery management system (BMS) is used, which acts based on the battery model. In past, equivalent circuit models (ECM) were employed, estimating simple battery state and parameters (e.g. ohmic resistance, open circuit voltage) using fixed safety constraints (e.g. fixed cut-off and cut-in voltage). This rendered simpler computation and easy implementation. But these models have limitations in providing insights into the internal behavior of the batteries (e.g. internal temperature, state of charge). Knowledge about electrochemical behavior is important because of two major reasons - as the battery ages, the operability regions changes (irreversible degradation due to loss of ions, active material, or electrolyte)., thus the fixed safety constraints may not guarantee safer and optimal operation. Secondly, newer automotive technologies demand broader operating conditions (e.g. fast charging, wider temperature profiles, etc.) which calls for advanced algorithms.

Electrochemical models (EM) play a major role in the future BMS. EM's are based on physiochemical mechanisms, resulting in a more accurate prediction while describing the internal states of the battery at wider operating ranges. However, the governing equations are non-linear and have multiple partial differential equations (PDE) that make them computationally expensive. Moreover, a large number of physical parameters are required which are often dependent on composition and temperature (e.g. thermodynamic potential, ionic conductivity, diffusion coefficient, transference number). Estimating these parameters with achievable accuracy continues to be a challenging job.

This study addresses two pivotal questions – how can we reduce or transform the complex full order EM maintaining the same accuracy at a wider operating window? We explore the suitability of different methods to convert PDE's (e.g. Finite Element Method, Pade approximation, Orthogonal decomposition, etc.) into simpler equations. Secondly, how to systematically estimate parameters without information loss? Parameter estimation techniques (e.g. Fisher Information Matrix, Particle Swarm Optimization, Genetic Algorithm) are studied to characterize the battery degradation and shed light on the cause-effect mechanisms.

Regression analysis and machine learning for the prediction of fire property damage based on fire type, weather conditions and site response characteristics

In modern society, along with the promotion of urbanization and industrialization due to the increase in population, the factors affecting the occurrence of fires are also diverse and complex with the development of science and technology. In cities with high population density due to urbanization and industrialization, different fire occurrence characteristics are shown depending on the use of buildings such as residential, industrial, and multi-use facilities, the structure of the building, and the size of the building. In addition, with the development of science and technology, technical fire factors such as mechanical and electrical factors and gas leakage are forming a complex relationship. Fire has been steadily occurring, and an average of about 42,000 fires have occurred annually over the past decade in Korea, and property damage and human casualties have also been steadily increasing. Fire statistics yearbooks are published every year, but as fires occur through complex relationships between various factors, the need for fire research on new concepts using the 4th industrial revolution technology to analyze and predict fire occurrence factors beyond conventional statistics is increasing.

This study aims to predict the size of fire damage by regression analysis of the characteristics of fire factors, and to continuously reduce damage to property and human life by preparing effective preventive measures based on quantitative models. Data were collected from the National Fire Data System (NFDS), and variables affecting property damage were selected as fire site arrival time, suppression time, ignition heat source, ignition factor, location, weather, building structure, etc. Prior to machine learning-based predictive modeling, SPSS, a statistical analysis program, was used to analyze the correlation between variables. In particular, linear and non-linear regression analysis were performed to create a predictive model according to fire type and suppression time to identify the increase in property damage. Then, the analyzed statistical data was applied to a machine learning-based predictive model and an improved predictive model was completed through training data learning.

Resolving a PEM Fuel Cell Dichotomy: Maximum-Power-Point or Maximum-Efficiency-Point Tracking?

Zuhair A. Alyousef

Department of Chemical Engineering, King Fahd University of Petroleum and Minerals

Dhahran 31261, Saudi Arabia

zyousef@kfupm.edu.sa

Oscar D. Crisalle

Department of Chemical Engineering, University of Florida,

Gainesville, FL 32611-6005, USA

crisalle@che.ufl.edu

Keywords: fuel cell, PEM fuel cell, DMFC, portable power

Abstract

Performance metrics for maximum-power-point tracking (MPPT) and maximum-efficiency-point tracking (MEPT) strategies for the management of open-cathode direct methanol fuel cells are characterized via simulation studies. The *liquid saturation level* is invoked for the novel role of serving as a performance metric that diagnoses potential failures. MPPT improves the fuel efficiency of the device; however, it suffers from significant disadvantages, including (*i*) lowering the voltage efficiency and thus the quality of electrical work, (*ii*) making the device vulnerable to failure by falling to a zero-voltage operating point under small perturbations, and (*iii*) driving the device perilously close to failure points due to the high production of water leading to saturation levels in excess of 90 % inside the pores of the liquid distribution layer used for oxygen molecules to reach the cathode reaction sites. In contrast, MEPT leads to (*i*) improvements in the real efficiency of the fuel cell by 18 %, (*ii*) safer operation away from limiting currents, (*iii*) better voltage efficiency, and (*iv*) maintaining an adequate liquid saturation level of 52 % for water in the pores leading to a significant mitigation of the risk of failure caused by cathode dry-out or flooding conditions.

1 Introduction

Direct methanol fuel cells (DMFCs) are appealing electrical power generation devices that function at low temperatures. Due to the small footprint of the devices that results from the use of liquid methanol, which is a high energy-density fuel, DMFCs stand as an attractive technology for power generation in portable electronic applications [1, 2, 3]. A DMFC can be viewed as an electrochemical reactor, where at the anode methanol and water undergo the *oxidation* reaction

$$CH_3OH + H_2O \rightarrow CO_2 + 6H^+ + 6e^- \quad (1)$$

that releases a stream of electrons in the form of an electrical current I that passes through an outer circuit, and in this fashion generating electrical power. The product H^+ protons migrate through a membrane positioned between the anode and the cathode, and upon reaching the cathode join the returning electrons as well as an oxygen stream to undergo the *reduction* reaction

$$6H^+ + 6e^- + \frac{3}{2}O_2 \rightarrow 3H_2O \tag{2}$$

Some methanol molecules diffuse through the membrane until they reach the cathode side, where they participate in the undesirable methanol-crossover reaction

$$CH_3OH + \frac{3}{2}O_2 \rightarrow 2H_2O + CO_2 \tag{3}$$

which reduces the fuel efficiency of the device because this form of methanol consumption does not produce free electrons needed for generating an electrical current.

It is critically important to manage the electrical current that adjusts the power generated by the fuel cell, because the objective of the device is to satisfy changing power demands of a connected load. Therefore, a robust methodology for specifying adequate operational regimes is necessary for the dual purposes of satisfying the power demand and preventing power shutdown failures [4]. Among the techniques gaining increasing literature interest is the maximum-power-point tracking (MPPT) method, which seeks to identify operating conditions that produce power at a peak value.

The fuel-cell literature includes a relatively modest number of investigated MPPT algorithms. For example, the work of Abdi *et al.* [5] on hybrid Proton Exchange Membrane Fuel Cell (PEMFC) systems implements MPPT in the context of a sliding-mode control strategy. Zhin-dan *et al.* [6] use an adaptive maximum-seeking algorithm in PEMFCs to estimate the maximum-power point in the outer loop of a real-time cascade control scheme. Restricting the focus to DMFCs, the work of Zhu *et al.* [7] suggests that MPPT may be interpreted as a quasimaximum-efficiency-point tracking method, after recognizing that the resulting operating points for maximum power and maximum efficiency are close to each other in the particular case of the DMFC architecture investigated in that work. However, the results in [7] are possibly dependent on design details of the DMFC system considered, and it would therefore be of value to the literature to investigate whether the conclusions and observations remain valid for a different class of DMFC architectures.

Maximum-efficiency-point tracking (MEPT) is an alternative operation management strategy. It has received less attention than MPPT in the fuel-cell literature. A significant precedent is the work of Vega *et al.* [8], where MEPT is implemented using a perturb-and-observe algorithm specialized for hydrogen fuel cells. Lu *et al.* [9] also focus on hydrogen fuel cell applications, relying on an extremum-seeking MEPT algorithm. A comprehensive comparative analysis work is presented by Kumaraswamy [10], where MPPT and MEPT methods are compared and contrasted for the case of hydrogen fuel cells. Kumaraswamy advocates discarding both the MPPT and MEPT techniques in favor of the adoption of a method that tracks a midpoint between the maximum power and the maximum efficiency points, as a tradeoff to mitigate the disadvantages associated with the other two methods. The authors have not found literature precedents that investigate MEPT techniques applied to open-cathode DMFCs.

The design of MPPT and the MEPT control algorithms involve two complementary operations. The first one is to select an MPP or MEP value as the *set point* for the controller. The second is the development and real-time deployment of an algorithm that adjusts the manipulated variables, which most typically include the electrical current, as needed to keep the operational state at the selected maximum point (a *regulation* control mode), or to move the current state to a new user-specified maximum-point state (*tracking* mode). This paper presents a comparison and contrast study of the merits of selecting a maximum-power or a maximum-efficiency operating point as the controller's set point. The tracking or regulation aspects control design are excluded from the scope of the work.

2 Performance Characterization and Results

This section presents an analysis methodology for device performance through significant metrics associated with the stack of fuels, as they relate to the Maximum-Power and Maximum-Efficiency points of operation. More specifically, the amount of electrical power generated and the real-efficiency of the fuel cell stack are selected the indicative operational performance criteria.

2.1 Device Performance Metrics

The output power of the DMFC is defined as

$$P = IE^{Stack} \tag{4}$$

and the *real efficiency* of the device is given by [11]

$$\varepsilon_{real} = \varepsilon_{thermo}\, \varepsilon_{voltage}\, \varepsilon_{fuel} \tag{5}$$

where *thermodynamic reversible efficiency* $\varepsilon_{thermo} = \triangle G / \triangle H$ depends on the change of Gibbs energy $\triangle G$ and on the heat of reaction $\triangle H$ of the electrochemical reaction. The remaining two efficiencies appearing on the right-hand side of Equation (5) are defined in Section 2.2. Given that the supply of energy to service a load is a primary objective of fuel-cell use, it is natural to consider the output power and the real efficiency as relevant metrics for DMFC performance assessment.

A flooding phenomenon occurring in open-cathode DMFCs stands as a most critical operational challenge for these devices [12, 13, 2]. Though a fraction of the water produced by the electrochemical reaction at the heart of all fuel cells is discharged to the ambient through an evaporation mechanism, at high electrical currents the evaporative water release rates can become very much smaller than the rate of liquid water generation by chemical reaction. Excessive water production causes a significant extent of liquid water invasion of the pores that serve as conduits for the oxygen that is needed by the device to produce electrical power. Eventually the level of flooding is such that the liquid water blocks the transport of oxygen into the device, and the fuel cell stops producing electrical power. This catastrophic operational failure is an undesirable degradation of device performance. In the DMFC architecture, the porous layer on the cathode side that holds and redistributes the electrochemically produced water, is known as the *liquid distribution layer* (LDL). The extent of water filling of the pore space can be quantified in terms of the *liquid saturation level* s_ℓ^{LDL}, defined as the ratio of the volume occupied by liquid water to the total free pore volume [14, 15]. This indicator quantifies the extent of pore flooding, adopting the limiting value of 100 % when the pore volume is completely filled with water, and 0 % when the pore volume has no water. In this work we calculate liquid saturation levels using the well-established Leverette-function approach for quantifying transport inside pores using experimentally gathered data. Details are available in the literature [16, 17, 15, 2].

This paper analyzes the MPPT and MEPT philosophies from the fuel-stack perspective, in a fashion that is valid independent of other components of the system. More specifically, other balance-of-plant power-consumption parasitic components (fuel injection pump, fans, etc.) are excluded from the analysis, as their specifications vary widely depending on preferred engineering-design criteria and are application specific. This permits the results to be applicable to all fuel-cell systems with electrochemical stacks whose polarization curves are qualitatively analogous to those shown in Figure1. Specific references to observations and results obtained in a DMFC system are presented for the purpose of providing supporting experimental and modeling evidence for the claims and conclusions of this work for at least one specific platform.

2.2 Maximum-Power Point Tracking

A maximum-power-point tracking control algorithm adjusts the manipulated variables of the fuel cell as needed to ensure that the device operates at a user-specified maximum-power-point value as the set point. More specifically an MPPT algorithm seeks to find an operating point, defined by a current I, voltage E^{Stack}, and temperature in the anode flow channel T^{AFC}, that together fix the output-power P of the device to its maximum value on the current-power polarization curve at temperature T^{AFC}. This section characterizes significant advantages and disadvantages that are inherent to the choice of an MPP as the set point for a tracking controller.

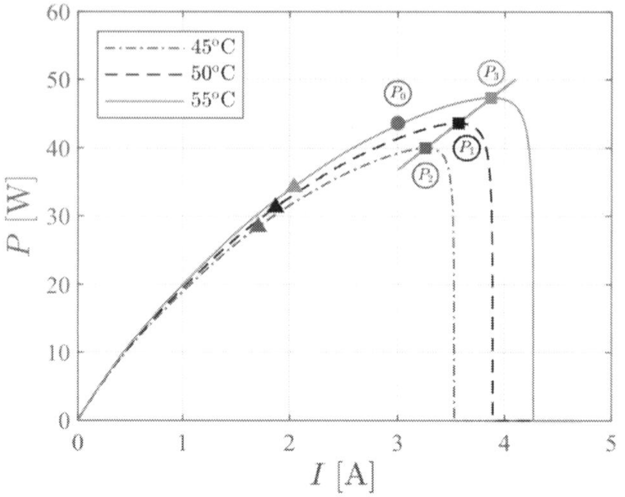

Figure 1: Isothermal current-power polarization curves at three temperatures. The square and triangle markers respectively identify the maximum-power point and the maximum-efficiency point at each isotherm. The straight-line segment connecting the markers defines the location of all possible maximum-power points located between the highest and lowest temperatures of operation acceptable for the device. The circled labels represent operating points discussed in Section 2.2.1.

Figure 1 illustrate the general principle of MPPT operation, as follows. The points labeled P_0 and P_1 align with the 43 Wvalue on the vertical axis. Hence, 43 W of power can be generated by operating at point P_0 (on the 55 °C isotherm) or point P_1 (on the 50 °C isotherm), because both points lie on the constant-power horizontal line passing through the 43 Wpower coordinate. Considering now a line segment on the 43 Whorizontal power line, with end points defined by both intersections of the line with the 55 °C isotherm, it follows that of all the points that lie on the segment only P_1 lies on the peak of an isotherm, and hence P_1 is designated as the *maximum-power point* (MPP) of operation at an output power of 43 W. The current-power coordinates of MPP P_1 on the figure are (3.57 A, 43 W). Consequently, to satisfy a power load demand of 43 W under an MPPT control strategy, the controller set point is set to an operating temperature of 50 °C while drawing an electrical current of 3.57 A.

2.2.1 MPPT Advantages and Disadvantages

A reported advantage of MPPT is a reduction of methanol consumption via the suppression of the undesirable crossover reaction (2), which consumes fuel without producing electrical power. Furthermore, operation at a maximum-power point has been reported to produce an increase in the *fuel efficiency* [18] of the device, defined as

$$\varepsilon_{fuel} = \frac{I}{I + I_x} \tag{6}$$

where the crossover current I_x represents the current loss that is incurred as a consequence of the crossover reaction (3). MPPT leads to higher fuel efficiency, as can be concluded by noting that, relative to P_0, MPP P_1 requires operation at a lower temperature (which reduces the cross over current) and at a higher current. The combined effect of these two changes at point P_1 produces a larger numerator and a smaller denominator in Equation (6), and hence leads to higher fuel efficiency.

One MPPT disadvantage follows from the precipitous fall in power observed in Figure 1 at locations close to each maximum-power point. The current where the rapidly falling isotherm reaches zero voltage is known as the *limiting current* of the isotherm [11], a point at which the device ceases to operate. Note, for example, that the current value of 3.57 A corresponding to MPP P_1 is in very close proximity of its isotherm's limiting current of 3.8 A. Hence, consistent operation at any maximum-power point requires currents in the near proximity of the catastrophic limiting-current point. This makes the MPPT method highly fragile, as it is vulnerable to perturbations that, by causing a slight increase in current, can induce a fatal operational failure.

A second disadvantage of MPPT is a potentially significant loss of *voltage efficiency* [11]

$$\varepsilon_{voltage} = \frac{E^{Stack}}{E_{Thermo}} \tag{7}$$

where E_{thermo} is thermodynamic voltage of the cell interpreted as theoretical maximum-possible voltage. The loss occurs because all MPPs are associated with currents that are very near the limiting current for their isotherms, a point where there is a sudden drop in power and therefore also in voltage. Low voltage efficiencies arise as the sharp drop in voltage that occurs next to all MPPs, decreasing the numerator in Equation (7).

A final disadvantage of MPPT is that all MPPs occur at high currents. As high currents are realized through higher rates of electron production by the electrochemical reduction reaction (2), it follows that the generation of liquid water, also a reaction product, concurrently experiences a rate increase. As a consequence, liquid water accumulates inside the cathode-side pores that act as supply channels for the oxygen needed to sustain the chemical reaction. This high-current scenario can easily induce a flooding state in the pores, where liquid water blocks the transport of oxygen, causing the cathodic reaction to stop as it is starved of oxygen.

2.3 Maximum Efficiency Point Tracking

In this case the manipulated variables are adjusted to ensure that operation is maintained at conditions that ensure that the output power is produced at the maximum-possible efficiency for a given operating temperature. From an engineering viewpoint, the most meaningful efficiency to choose as the optimization objective is the real efficiency, defined in Equation (5), because it simultaneously addresses all three major efficiencies of relevance to device performance.

2.3.1 MEPT Advantages and Disadvantages

This work identifies advantages of the MEPT approach that have not been reported in prior literature. A discussion is given in the next section. As an important contextual reference, in this section we point out a potential disadvantage of the technique. The shortcoming derives that, in contrast to maximum-power points, maximum-efficiency points may be difficult to identify from the available measurements. The quantification of MEPs may require measurements of methanol concentration at the inlet of the anode flow channel (a typically available measurement) and at the outlet point (a measurement that is not usually available because it requires a second concentration sensor).

2.4 Results

Figures 2(a) and (b) include three graphs, respectively showing the performance results obtained under the MPPT and MEPT fuel-cell management techniques at three temperatures, as a function of the electrical current. The top graph in each plot triplet shows the real efficiency ε_{real}, the middle graph is for the stack voltage E^{Stack}, and the bottom graph represents the liquid saturation level s_l^{LDL}. All markers on Figure 2(a) correspond, at matching current values, to the square MPP makers in Figure 1, and all markers in Figure 2(b) correspond to the triangle MEP markers in Figure 1.

Figure 2(a) characterizes the performance of the MPPT technique. The MPP markers on the top graph indicate that MPPT delivers real efficiencies of approximately 21.5 % at the 45, 50, and 55 °C operating temperatures. The middle graph shows that all MPPs occur at currents that are very close to the zero-voltage current, where power generation shuts down. Finally, the MPPs in the bottom plot correspond to very large liquid saturation levels, in

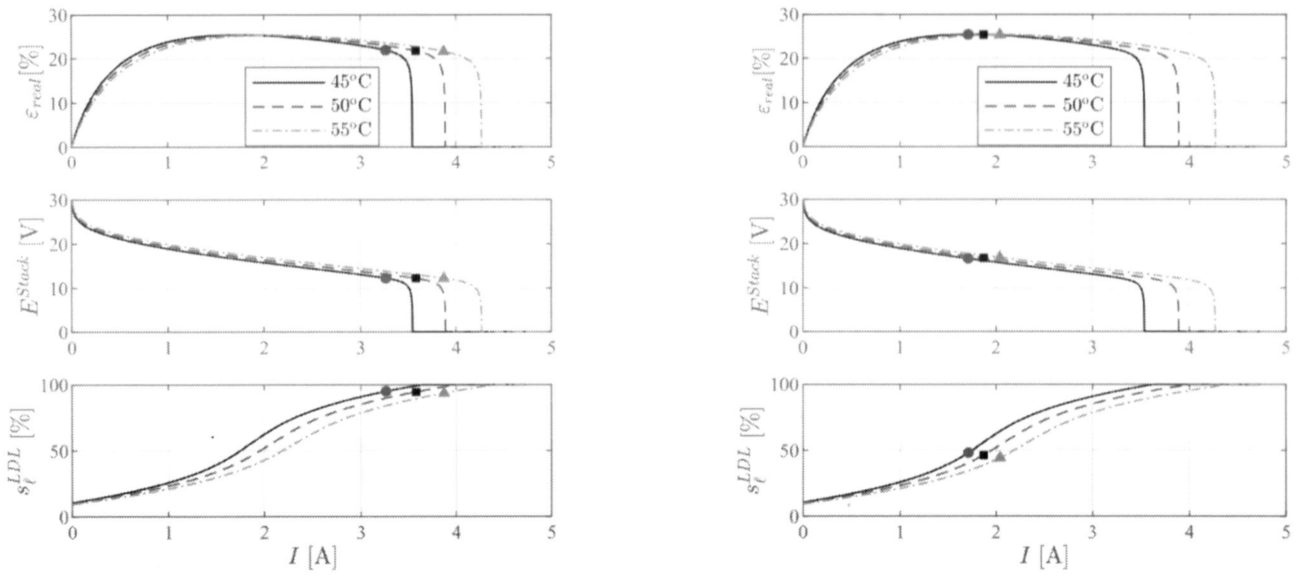

(a) Markers identifying the maximum-power points on three isotherms.

(b) Markers identifying the maximum-efficiency points on three isotherms.

Figure 2: Real efficiency ε_{real}, stack voltage E^{Stack}, and liquid saturation level s_ℓ^{LDL} in the liquid distribution layer on three isotherms as a function of the electrical current I.

excess of 90 %. Hence, less than 10 % of the pore volume is free of liquid water at any MPP, which is a small amount of pore space to support effective oxygen diffusion.

Figure 2(b) characterizes the performance of the MEPT technique. The top graph shows that the maximum real efficiency is approximately 25.3 % at all temperatures, surpassing the 21.5 % MPP efficiency. At MEPs zone, the difference between these efficiency values is insignificant for all practical purposes, and they are located in proximity to each other over a narrow range of electrical currents. The middle graph shows that at all temperatures the MEPs are obtained at current values that are significantly distant from their corresponding limiting currents. More precisely, all MEPs are at least 1.8 A away from the catastrophic limiting current threshold for their isotherm. The stack voltage for all MEPs is 16.5 V on average, a significantly higher voltage than the 12 V realized at MPPs. Finally, the bottom graph reveals the most surprising insight, namely that at all MEPs the liquid saturation levels in the cathode-side pores never exceed 48 %, leaving a significantly larger liquid-free volume for oxygen transport than at any MPP. Flooding failures are unlikely because so much of the pore volume does not contain water. Furthermore, approximately 52 % of the pore still contains liquid water, which in turn makes it unlikely that the device will shut down due to the abrupt failure mode that occurs when the pores and cathode dry out, stopping all electrochemical reactions.

In summary, the flooding analysis results point out three additional advantages of MEPT over MPPT operation beyond improvement of the real efficiency. A first advantage is that the MEP currents are fairly distant from limiting current values, in contrast to the MPPs which lie perilously close to flooding currents that cause shutdown. Second, at all temperatures the voltage losses under a maximum-efficiency regime are significantly smaller than the corresponding losses in a maximum-power operating point. As a consequence, the voltage efficiency defined in Equation (7) is higher under MEPT than under MPPT operation. The results show voltage efficiencies at values that are 38 % higher at MEPs than at MPPs. Finally, a third additional advantage is that, at any temperature, the problem of power failure by cathode flooding or drying is mitigated to a very large extent by operating at MEP conditions rather than at MPP conditions.

3 Conclusions

The performance comparison of the flooding regimes under the MPPT and the MEPT methods resulting from the analysis of Figures 2(a) and (b) given in Section 2.4, demonstrates that the MEPT method is superior not only in terms of delivering the best-possible real efficiency, but also by keeping the cathode-side pores significantly away from a flooding limit and operating at currents that are significantly lower than the forbidden upper bound imposed by the limiting current. This work addresses the principles of the MEPT approach from a set-point perspective but does not include in its scope an attempt to design a tracking or regulation controller, a task that is left as a proposed future work.

References

[1] Biswas M, Mudiraj S, Lear W, Crisalle O. Systematic approach for modeling methanol mass transport on the anode side of direct methanol fuel cells. *International Journal of Hydrogen Energy*. 2014;.

[2] Mudiraj SP. Advanced Model-based Control Design for Direct Methanol Fuel Cell Systems. phdthesis, University of Florida. 2015.

[3] Zenith F, Krewer U. Modelling, dynamics and control of a portable DMFC system. *Journal of Process Control*. 2010; 20:630–642.

[4] Lauzze KC, Chmielewski DJ. Power control of a polymer electrolyte membrane fuel cell. *Industrial & engineering chemistry research*. 2006;45(13):4661–4670.

[5] Abdi S, Afshar K, Bigdeli N, Ahmadi S. A novel approach for robust maximum power point tracking of PEM fuel cell generator using sliding mode control approach. *International Journal of Electrochemical Science*. 2012;7(1):4192–4209.

[6] Zhi-dan Z, Hai-bo H, Xin-jian Z, Guang-yi C, Yuan R. Adaptive Maximum Power Point Tracking Control of Fuel Cell Power Plants. *Journal of Power Sources*. 2008;176:259–269.

[7] Zhu GR, Loo K, Lai Y, Chi KT. Quasi-maximum efficiency point tracking for direct methanol fuel cell in DMFC/supercapacitor hybrid energy system. *IEEE Transactions on Energy Conversion*. 2012;27(3):561–571.

[8] Vega DH, Marx N, Boulon L, Hernandez A. Maximum efficiency point tracking for hydrogen fuel cells. In: *2014 IEEE 27th Canadian Conference on Electrical and Computer Engineering (CCECE)*. IEEE. 2014; pp. 1–6.

[9] Lu J, Zahedi A. Maximum efficiency point tracking control for fuel cell power systems. In: *2010 International Conference on Power System Technology*. IEEE. 2010; pp. 1–6.

[10] Kumaraswamy VK. A standalone proton exchange membrane fuel cell generation system with different tracking techniques. Ph.D. thesis, Memorial University of Newfoundland. 2017.

[11] O'Hayre R, Cha S, Collela W, Prinz F. *Fuel Cell Fundamentals*. John Wiley & Sons, 2nd ed. 2009.

[12] Oliveira V, Rangel C, Pinto A. Water Management in Direct Methanol Fuel Cells. *Internation Journal of Hydrogen Energy*. 2009;34:8245 – 8256.

[13] Zago M, Casalegno A, Santoro C, Marchesi R. Water transport and flooding in DMFC: Experimental and modeling analyses. *Journal of Power Sources*. 2012;217:381–391.

[14] Mossman A, Wells B, Barton R, Voss H. Passive recovery of liquid water produced by fuel cells. 2012. US Patent 8,298,719.

[15] Mudiraj S, Biswas M, Lear W, Crisalle O. Comprehensive mass transport modeling technique for the cathode side of an open-cathode direct methanol fuel cell. *International Journal of Hydrogen Energy*. 2015;40:8137–8159.

[16] Ye Q, Van Nguyen T. Three-dimensional simulation of liquid water distribution in a PEMFC with experimentally measured capillary functions. *Journal of the Electrochemical Society*. 2007;154(12):B1242–B1251.

[17] Pasaogullari U, Wang C. Liquid water transport in gas diffusion layer of polymer electrolyte fuel cells. *Journal of the Electrochemical Society*. 2004;151(3):A399–A406.

[18] Chippar P, Ko J, Ju H. A global transient, one-dimensional, two-phase model for direct methanol fuel cells (DMFCs)–Part II: Analysis of the time-dependent thermal behavior of DMFCs. *Energy*. 2010;35(5):2301–2308.

639636 Automate More Than Your Processes, Automate Your Procedures

Abstract:

By: Brian Bailey (and/or David Wicker) of ProLytX and Chris Reeves of Koch Ag & Energy Solutions

Problem / Scope

Safety Instrumented System (SIS) testing is a US regulatory requirement, but the quality, coverage, repeatability, and auditability vary widely from site to site. Often, schedule and budget concerns or staffing limitations pressure owners to trim tests down, focusing on speed of execution rather than quality validation. In many cases, even a quality test when done quickly can be hard to audit.

Koch Ag & Energy Solutions (KAES), in partnership with ProLytX, has moved towards a better solution. The key to this solution is adopting a segmented testing approach and combining it with automated validation software to dramatically reduce plant downtime and cost while improving their facilities' overall quality of testing and safety.

Methodology

IEC61511 allows for segmented testing (testing in parts). Using this methodology, all SIS logic can be validated outside of a Turnaround or maintenance window and coupled with loop tests and field inspection to provide full coverage. KAES and ProLytX will present how a combination of automated testing software and segmented testing methodology can reduce downtime and manpower while also improving test coverage and documentation.

Results

Data from multiple sites (100+) will be shown that document plant savings and testing discoveries. Critical path reductions typically land between 0.5 to 5 days with immediate ROI on procedure development. The presentation will cover sample test failures, logistics, comparison of manual and automated methodology (pros and cons), and suggested modifications to site testing and Management of Change (MOC) requirements.

Implications

Reduce human errors and execution time through automation and segmented testing.

Improve testing quality and coverage with extensive procedures.

Improve safety by finding hidden failures and testing abnormal conditions.

Background

Brian Bailey, CFSE – Director of Functional Safety @ ProLytX - ~17 years in Automation and SIS Implementation.

David Wicker – Director of Technology @ ProLytX - ~17 years in Automation and SIS Implementation.

Chris Reeves – Functional Safety Manager / Reliability SME - ~30 years in Instrumentation / Automation*

*Koch Ag & Energy Solutions details to be vetted by legal prior to submittal. Do not share abstract details outside of the organization until the paper is approved.

639337 Numerical Model to Evaluate the Consequences of a Gas Release from a Buried Pipeline: Model Construction, Validation and Preliminary Results

Abstract:

Depending on the flow and soil characteristics, the underground gas releases from buried pipelines result in different regimes. The regimes range from migration through the soil, to a visible uplift on the ground, to a crater formation. The proper identification of the regime is key to an accurate quantification of the consequences; thus, a successful risk assessment. There is available work that discussed each regime separately; however, there is not a continuous model that identifies all kinds of regimes simultaneously and the transition between them. In this study, we present a universal computational model, with simplified resources, that covers the whole range of regimes. We developed the model based on the Eulerian approach for fluid flow and the kinetic theory of granular flow for soil as a granular phase. Moreover, we established quantitative guidelines serving as a mechanistic approach to identify the regime. Using the model, we performed a sensitivity analysis to test the effect of some parameters on the regime. This analysis revealed that a higher pressure, a lower pipeline burial depth, and an upper release orientation trigger more damaging consequences. Furthermore, the results from this analysis were presented on a nomograph that allows the determination of the regime for a given release scenario of pipeline pressure and burial depth.

639402 Gas Pipeline System Integrity Management Software Platform Based on Dynamic Corrosion Bayesian Network Predictive Modeling

Abstract:

Pipeline failures are recurring over the last decades and are causing not only property damage but human injuries and lives as well. Improving the current pipeline integrity tools and techniques is becoming the main focus of operators around the world. Quantifying the health state of the pipeline system is the main challenge facing the operators due to the limited sensor monitoring points (limited budget) which can provide only partial health state data and non optimal maintenance schedules. As a result, the challenge is to find the optimal locations to place sensors to capture the health state of the pipeline system in the most accurate way possible and optimal inspection/maintenance schedules. This paper presents a software platform for gas pipeline system integrity management which includes novel sensor placement optimization framework that aims at providing the operators the optimal locations to place sensors at a minimal cost and the optimal mitigation actions and their timing. This framework maximizes the damage detection probability by the provided sensor network which captures as much as possible of the degrading parts of the pipeline. The proposed real-time optimization framework is based on dynamic corrosion Bayesian Network modeling where the changing operating parameters are updating the system simulation for optimal results. The real-time sensor placement optimization will provide the optimal sensor network layout and capture damages that could have been gone undetected and resulted in pipeline failures. In a few words, the proposed research work will help operators make risk-informed decisions to enhance pipeline system integrity management through optimal mitigation actions.

639192 Next-Generation Simulation Platform for Sustainable Process Design and Operation

Abstract:

Global sustainability and net zero imperatives are driving process industries to embrace tools to reduce emissions and improve their overall process sustainability. The transition towards sustainability will be implemented through near-, medium-, and long-term initiatives that may require significant modification to long-held processes and substantial capital investment. In the near-term, companies will look towards optimization of existing processes to reduce carbon emissions. Medium- and long-term projects will focus on the switch to renewable energy, development of new products that meet the demand of the circular economy, and production processes with improved energy efficiency. As Owner-Operator and EPC companies prepare to make this transition, they will be reliant on process simulation tools to help drive innovation with accelerated schedules and limited capital.

Traditional process simulators are well-suited for modeling standard processes but will struggle to meet the demands of today's digital transformation and energy transition. Their conventional architectures mean they are not ideal for modeling emerging technologies in the hydrogen and low-carbon economy. Limited integration points also create challenges for engineers who require a Digital Twin for the entire lifecycle of a process.

This presentation focuses on how a next-generation simulation platform can address the challenges faced by traditional single-purpose simulation tools to meet these global sustainability imperatives. The new simulation platform should provide:

1. An open modelling approach to shorten the development time of new process equipment models for wind and solar power generation, water electrolysis, and biodiesel production.
2. Methods for calculating greenhouse gas emissions and sustainability metrics (energy intensity, carbon efficiency, water intensity, and others) to enable optimization of existing processes.
3. Integrated tools for multi-objective optimization of operating cost and process sustainability.
4. One simulation model for steady-state and dynamic process design for understanding the impact of variable wind and solar resources on the process.
5. Connectivity to process historians and distributed control systems for real-time plant optimization using an engineering Digital Twin.

This presentation will introduce AVEVA Process Simulation as an example of a simulation platform that can simultaneously model new sustainable processes and accelerate their engineering design.

639345 Efficient and Accurate Simulation of Stirred Mixing Tanks in a Wider Context

Abstract:

Stirred mixing tanks are one of the work horses in the chemical, process, food & beverages and pharma industry. Computational Fluid Dynamics (CFD) simulations are nowadays widely used to characterize mixing tanks as well as to obtain a deeper understanding of the underlaying mechanisms and physics. A broad variety of different methods so solve the describing equations are available for industrial use like Finite Volume (FV), Finite Element (FE), Smooth Particle Hydrodynaics (SPH) or Lattice-Boltzman (LB), to mention a few. For these methods numerous models were developed to describe the relevant physics, the motion of the impeller(s), turbulence, dispersed or stratified multiphase flow with and without reactions, but also to quantify engineering quantities efficiently like power input, mixing time or yield of a chemical reaction. Obviously, depending on the engineering question which has to be answered and depending on the context, different level of accuracy and computational runtime have to be met. One can easily imagine that in the detailed engineering phase of a plant the requirements are significantly different from the ones during operation.

In this contribution we will discuss the benefits of different methods and models w.r.t. the different phases of a plant life and how this fits into a wider simulation context via e.g. reduced order models. A special focus will be on turbulence modeling, which has a significant impact on accuracy, but also on computation times.

638979 Development of a Model Reduction Approach for a Pressure-Driven Column Model Using Digital Twin Technology

Abstract:

In recent years, an increase in the volatility of the German energy market was apparent caused by the rising share of renewable energies in the power supply. Thus, major energy consumers such as cryogenic air separation units (ASUs) need to adapt their operations to stabilize the power grid. The high potential of flexible ASU operation is investigated in the Kopernikus project "SynErgie - FlexASU" funded by the German Federal Ministry of Education and Research (BMBF).

A major aspect of efficient and flexible plant operation is the development of advanced control strategies. Kender et al. 2021 [1] introduced a digital twin of an ASU which contains a pressure-driven detailed dynamic plant model, the virtual ASU, as a core component.

The digital twin allows for the in-silico development of novel control strategies considering the whole operating range of the plant. However, the dynamic plant simulation requires a high computational effort and is therefore not suitable for real-time applications. To address this issue, the adapted edmister model (AEM) presented by Ecker et al. 2019 [2] is used as a basis to develop a model reduction approach for a pressure-driven column model. The resulting dynamic edmister model (DEM) fulfills the high requirements regarding robustness and simulation accuracy while reducing the computational time. In addition, it can be applied for arbitrary column configurations and is incorporated inherently into the virtual ASU to reduce its model granularity utilizing it for applications such as dynamic optimization and nonlinear model predictive control.

[1] Kender R., Kaufmann F., Rößler F., Wunderlich B., Golubev D., Thomas I., Ecker A. M., Rehfeldt S., Klein H.: Development of a Digital Twin for a Flexible Air Separation Unit Using a Pressure-Driven Simulation Approach. Computers & Chemical Engineering, 151, 107349, **2021**

[2] Ecker A. M., Thomas I., Häfele M., Wunderlich B., Obermeier A., Ferstl J., Klein H., Peschel A.: Development of a new column shortcut model and its application in process optimisation. Chemical Engineering Science, 196, 538-551, **2019**

639407 A Data-Driven Approach for Modeling and Energy Optimization of Smr Hydrogen Production Process

Abstract:

Recently, the attention of hydrogen as a clean energy source has increased, because it can deal with various issues related to the fossil fuel. The hydrogen is usually supplied by three primary methods and the steam methane reforming(SMR) process reforming the natural gas is the most common method. In the case of small scale of SMR process, the SMR reactor has a complex structure such as double-tube type of reactors. The complexity has limitations in expressing the systems with several assumptions. This study focuses on modeling the SMR process by data-driven approach and optimization for maximizing the thermal efficiency of the process using intelligent optimization algorithms. A deep neural network(DNN) has been constructed to find the relations between the process variables and trained by actual operation data. To improve the model accuracy, the noise and outliers in the data are removed by data preprocessing methods. The optimization of neural networks which are black box models is implemented in a heuristic way to find the optimal point of the system. However, the result highly depends on the type of optimization algorithm and it is important to select the most appropriate one for the system. Three optimization algorithms are used to search for the optimal solution: grid search(GS), genetic algorithm(GA), and particle swarm optimization(PSO). As a result, the GS shows the fastest speed to find the optimal solution among them because the it has no iterative calculations. Unfortunately, when the number of decision variables increases, the calculation cost has greatly increased and optimization is not available. On the other hand, the number of decision variables is not limited in the GA and PSO. The thermal efficiency of the process increases as the number of decision variable increases, but the search time increases significantly in GA. Therefore, the PSO shows the best performance to optimize the SMR process. Consequently, the optimal thermal efficiency of 85.4% is obtained and this result will contribute to increase the productivity of hydrogen.

639496 Estimation of the Binary Interaction Parameters of the Anrtl Model Using Molecular Simulations

Abstract:

The adsorption Non-Random Two-Liquid (aNRTL) model[1] captures the non-idealities in the adsorbed phase for mixed gas adsorption equilibria. The underlying premise of the model is based on the dominance of adsorbate-adsorbent interactions in the adsorbed phase. It regresses the binary interaction parameter which is used to calculate the adsorbed phase mole fractions and the activity coefficients. The binary interaction parameter is a measure of competitive adsorption of binary gas mixture. This model requires regression of model parameters using experimental data. Unavailability or unreliability of experimental datasets in terms of missing uncertainties and fewer data points affects the regression process, thus impacting the model output.

Therefore, it is extremely important to develop a methodology to estimate such parameters that does not make use of experimental adsorption data. Such an approach has been developed for the estimation of binary interaction parameters of the NRTL model using molecular dynamics simulations[2]. In this study, we follow a similar approach to estimate the binary interaction parameters for a pair of adsorbate molecules in zeolites and MOFs using Monte Carlo and Molecular Dynamics Simulations. Following arguments like the ones used by Brandini and Prausnitz (1982)[3] and Ravichandran et al. (2018)[2], we have derived equations relating the binary interaction parameters to the molecular parameters of the adsorbate-adsorbent interactions in the first coordination shell of each distinct adsorption site. The binary interaction parameters computed from molecular simulations gave an accurate prediction of the binary adsorption equilibria for the systems that were studied.

References:

1. Kaur, Harnoor, Hla Tun, Michael Sees, and Chau-Chyun Chen. "Local composition activity coefficient model for mixed-gas adsorption equilibria." *Adsorption* 25, no. 5 (2019): 951-964.
2. Ravichandran, Ashwin, Rajesh Khare, and Chau-Chyun Chen. "Predicting NRTL binary interaction parameters from molecular simulations." *AIChE Journal*64, no. 7 (2018): 2758-2769.
3. Brandani, V., and J. M. Prausnitz. "Two-fluid theory and thermodynamic properties of liquid mixtures: General theory." *Proceedings of the National Academy of Sciences*79, no. 14 (1982): 4506-4509.

639550 Static Adsorption Modelling of a Cationic Gemini Surfactant in Different Rocks

Abstract:

The need to minimize surfactant adsorption on rock surfaces has been a challenge in chemical enhanced oil recovery (cEOR). This requires an accurate understanding of the adsorption mechanism on rocks. The static adsorption of an in-house cationic gemini surfactant on different rocks has been studied in this paper. Two different classes of rocks have been studied, namely sandstones and carbonates. XRD and XRF studies were conducted for rock samples. Different types of sandstones have also been analyzed. Each type showed a similar trend but a different amount of surfactant adsorption. The major mechanism of gemini surfactant adsorption on rocks is the electrostatic attraction between the positive charges of the gemini surfactant and the negative sites present on the rock surface. Adsorption in sandstone rocks was found significantly higher than that in carbonates because of the high density of negative charges. The differences in the quantity of adsorped surfactant can be explained by different layer structures. It is proposed that an interdigitated bilayer (equivalent to a monolayer) is formed on carbonate rocks, whereas a non-interdigitated bilayer is on sandstone rock samples. The adsorption value at the plateau region is around 1 mg/g-rock in carbonates.

Modeling of adsorption experimental data helps in estimating the extent of adsorption and, hence, optimizing the process. We have applied curve-fitting to the experimental data using several adsorption isotherms, such as Langmuir, Freundlich, Redlich-Peterson, and Hill isotherms. The curve fitting was carried out using the Levenberg Marquardt algorithm to estimate the model parameters. The error metric consists of mean squared error and coefficient of determination. A better error metric was obtained using a Redlich-Peterson isotherm in sandstone, whereas both Langmuir and Redlich-Peterson isotherm performed better for carbonates. The modeling results reflect the proposed adsorption mechanism. This paper reports the adsorption mechanism of a novel cationic gemini surfactant in different types of rocks based on experimental data and adsorption isotherm modeling, which has great importance in cEOR.

Analysis of Two Multi-Phase Stirred Vessels Utilizing Computational Fluid Dynamics (Presentation #639908)

Ralph J. Price
Chevron Phillips Chemical
Kingwood, TX

John Thomas
M-Star Simulations
Friendship, MD

Abstract

A study was completed to determine potential root causes for performance differences between a pilot-plant stirred vessel and a commercial stirred vessel. Time-resolved computational fluid dynamics (CFD) simulations were utilized to determine the flow patterns and residence time distributions (RTDs) of fluid and different particle sizes in these two stirred vessels utilizing M-Star CFD software. This software is capable of predicting time-resolved turbulent fluid flow as well as particle dynamics within the flow. To determine RTDs, tracers (represented as massless particles) and 3 different particle sizes (small, medium, and large) were introduced into the inlet in an instantaneous pulse. Pilot-plant CFD results with constant inlet and outlet flows showed that the fluid RTD closely aligned with its corollary continuously stirred tank reactor (CSTR) RTD, but the particle RTDs varied significantly from the fluid RTD. The small particle size had a mean residence time ~3x the fluid mean residence time while the mean residence time of the large particle size could not be determined with any degree of certainty because so few particles exited over the computational run time. Actual pilot-plant operation is known to have a periodic pulsed outlet flow. An approximation of this pulse was applied as a transient boundary condition. When compared with the constant outlet flow simulation, the pulsed outlet flow simulation showed significant differences in the RTDs for all particle sizes, but only minor differences in the fluid RTD. The particle RTDs still varied significantly from the fluid, with the large particle size having a mean residence time ~55x the fluid's mean residence time. In contrast to the pilot-plant simulations, the commercial vessel simulation showed that the fluid RTD closely matched the CSTR RTD and the particle RTDs showed much less variation and closer agreement with the CSTR RTD, with the mean residence time of the large particle size ~2x the fluid mean residence time. These results indicate that the performance differences between the pilot plant and commercial scales are likely significantly impacted by the particle RTDs. These results also indicate the need for accurate time-resolved operational data for the pilot plant, especially the outlet flow pulse, to be able to accurately determine the RTDs of the different particle sizes utilizing a CFD model.

Introduction

Stirred vessels, including stirred reactors, are important equipment in a significant number of industrial and laboratory chemical processes. Stirred vessels can include batch or continuous processes, and can involve a single phase or multiple phases (liquid/gas, liquid/liquid, liquid/solid). For the system of interest, particles and liquid are continuously mixed in a vessel with a given residence time. Performance differences between pilot scale and commercial scale vessels have been noted since the commercial vessel was commissioned. These vessels were designed to provide approximately equal mean residence times. The mean residence time was computed using the ideal CSTR residence time formula shown in Equation 1, where τ is the mean residence time, V is the reactor volume, and v_0 is the inlet volumetric flow rate.

$$\tau = \frac{V}{v_0} \qquad \textbf{Equation 1}$$

A hypothesis for the difference in performance was differences in the residence time distributions in both vessels, particularly the impacts of inadequate mixing in the commercial stirred vessel. CFD modeling was performed to determine the validity of this hypothesis.

Results and Discussion

Initially, a steady-state simulation was used to evaluate the RTDs for particles in both reactors. In that study, a steady-state velocity profile was predicted and then around 100 particles (99 for the pilot-plant stirred vessel and 67 for the commercial stirred vessel) were injected into this converged flow field to calculate particle paths and RTDs. The steady-state flow field was held constant under the assumption that the particles do not affect the flow field. The results of the work were inconclusive, with the majority of all three particle sizes predicted to remain in the pilot-plant stirred vessel for a significant time (~1 mean residence time), shown in Table 1. However, during operation, only slow accumulation of particles occurs in the pilot-plant stirred vessel, so there is a disconnect between modeling and operation.

Table 1. Residence Time Distributions Predicted Utilizing a Steady-State CFD Model

Description	Relative Particle Size	Particles Exiting Pilot Plant in a Mean Residence Time	Particles Exiting Commercial Plant in a Mean Residence Time
Small	1.0	1 (1%)	35 (52%)
Medium	1.9	0 (0%)	7 (10.4%)
Large	3.4	0 (0%)	1 (1.5%)

Because these initial steady-state simulations were inconclusive, both stirred vessels were modeled using M-Star to ascertain any dynamic flow phenomena that impacts performance. These simulations had the same constant inlet flows and stirrer rotational speed as the steady-state model. Comparing the velocity fields between the Fluent steady-state simulation (left graphic in Figure 1) and the M-Star time-averaged simulation (middle graphic of Figure 1) for the pilot plant stirred vessel shows fairly similar velocity results between these two models, representing general agreement on the overall time-average fluid motion. However, small turbulent eddies, seen in the right graphic of Figure 1, represent time-resolved behavior that influences individual particle paths in the stirred vessel. This is especially apparent in judging the difference between the peak time-resolved velocity (1.5 m/s) and the peak time-averaged velocity (0.77 m/s), as seen in Figure 1.

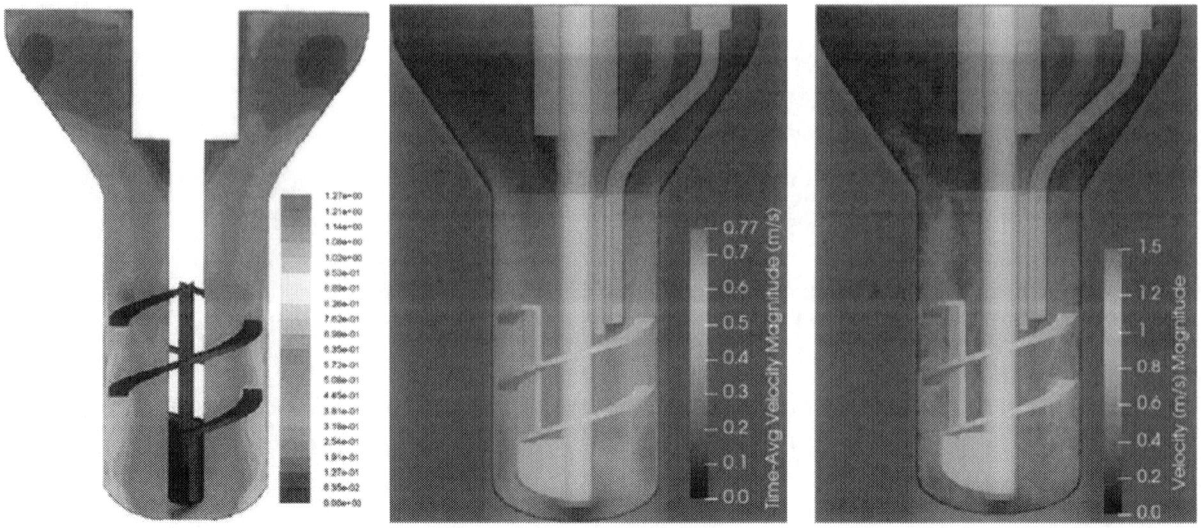

Steady State (Fluent) | Time-Averaged (M-Star) | Instantaneous (M-Star)

Figure 1. Pilot Plant Mixing Vessel Velocity Profiles

An RTD analysis for 3 different particle sizes was performed using the M-Star simulation. One million (1,000,000) particles of the 3 different sizes, along with the same number of massless tracer particles, were utilized to determine the RTDs for the small, medium, and large particle sizes and the reactor fluid. Compared to the mean residence time, only a relatively small time (5.8% of the mean residence time) was simulated. The combination of the simulated time with the large number of particles/tracers enables comparison of the CFD results with the CSTR equation, shown in Equation 2, where F is the fraction of fluid remaining that was injected at t=0, t is the current time, and τ is the mean residence time.

$$F(t) = 1 - \exp\left(\frac{-t}{\tau}\right) \qquad \textbf{Equation 2}$$

The RTD results of the pilot plant are shown in Figure 2. The RTD results of the commercial plant are shown in Figure 3, with the results from both stirred vessels summarized in Table 2.

For the pilot plant stirred vessel, the CFD results indicate that the particle RTDs are significantly different than the fluid RTD. The main reason for the longer residence times for particles is the relatively low exit velocity. The predicted exit velocity is less than the terminal settling velocities for all the particle sizes, although it is close to the settling velocity of the small particle size. Because of this, the main way particles exit the stirred vessel is in a turbulent eddy traveling into the exit piping. Most turbulent eddies in this system are created by the stirrer and dissipate in the upper section of the stirred vessel, so only a few eddies can carry particles out of the stirred vessel. In addition, the stirrer rotation imparts momentum to the particles, and, because the particles are denser than the fluid, their inertia causes them to preferentially segregate towards the outside radius of the vessel away from the vessel outlet, which is located towards the middle on the top of the reactor.

For the commercial stirred vessel, the results indicate that all but the large particle size behave somewhat similarly to the fluid, varying by less than 20%. The large particles are predicted to have a mean residence time of ~2x that of the fluid. Compared to the pilot plant stirred vessel, even the large particles do not significantly vary from the fluid. This is mostly due to more turbulence in the vessel from the stirrer design and the internal stationary baffles. In addition, the velocity in the outlet pipe is significantly above the terminal settling velocity for all the particle sizes.

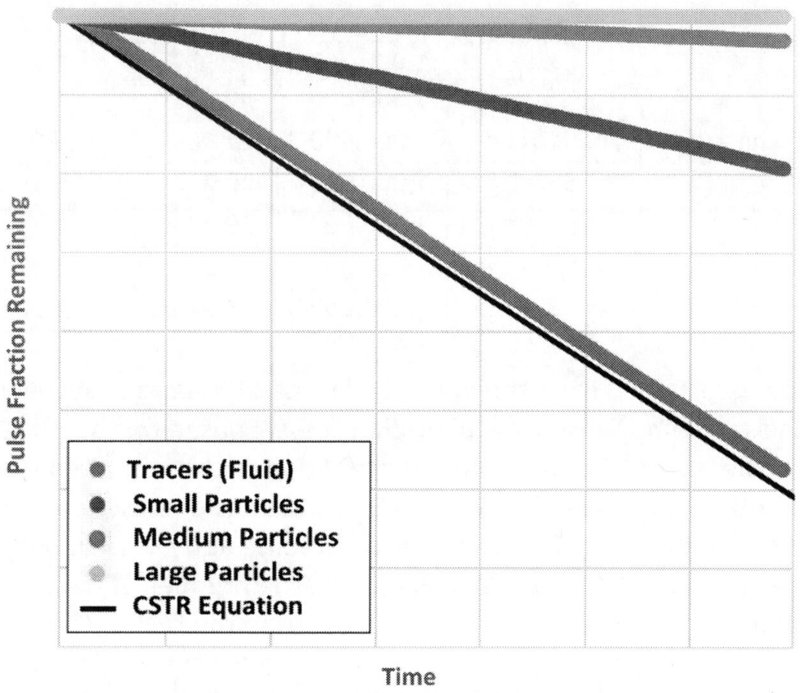

Figure 2. Pilot Plant Stirred Vessel RTDs of the 3 Different Particle Sizes and Tracers

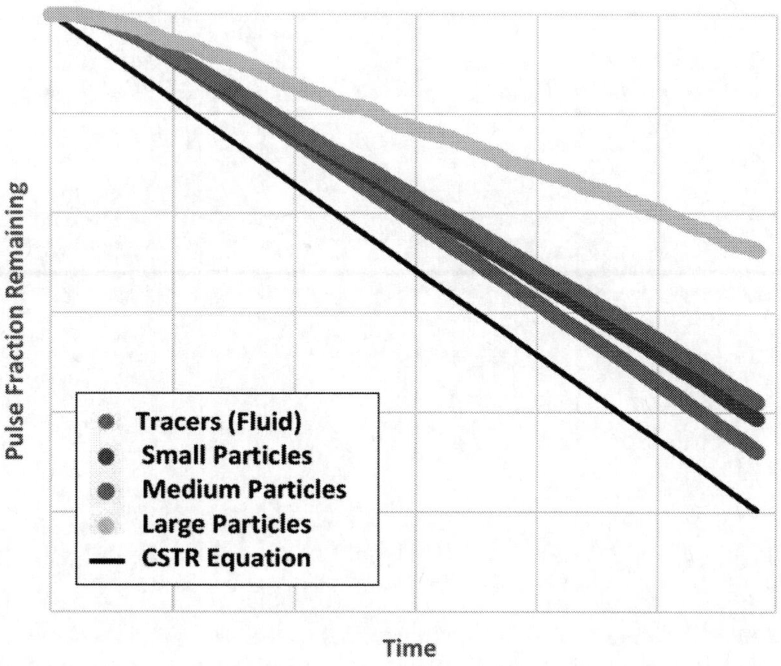

Figure 3. Commercial Stirred Vessel RTDs of the 3 Different Particle Sizes and Tracers

Table 2. RTDs and Mean Residence Times Predicted for both Stirred Vessels

Stirred vessel	Particle Description	Initial No.	No. Exited at End Time	Comparative Mean Res. Time
Pilot Plant	Fluid (tracer)	1,000,000	54,173	1.02
	Small	1,000,000	18,008	3.15
	Medium	1,000,000	2,843	20.4
	Large	1,000,000	9	NM
	CSTR	1,000,000	59,918	1.00
Commercial Plant	Fluid (tracer)	1,000,000	21,965	1.05
	Small	1,000,000	20,323	1.14
	Medium	1,000,000	19,476	1.18
	Large	1,000,000	11,852	1.97
	CSTR	1,000,000	24,920	1.00

Another conclusion ascertained from the commercial vessel RTDs is that the initial mixing time is greater for this vessel compared to the pilot-plant vessel based on the offset between the CSTR equation and the tracer RTD. This was expected due to the volume difference between the two vessels. Another conclusion from the CFD predictions of the commercial stirred vessel is that there is little fluid exchange in the cylindrical volume above the upper bell. This is seen in snapshots of the tracer particles shown in Figure 4. The number of tracers and particles in the top section of the stirred vessel is significantly less than the rest of the stirred vessel, indicating this volume is likely not fully utilized and should potentially be eliminated as active volume for any CSTR approximation of this stirred vessel.

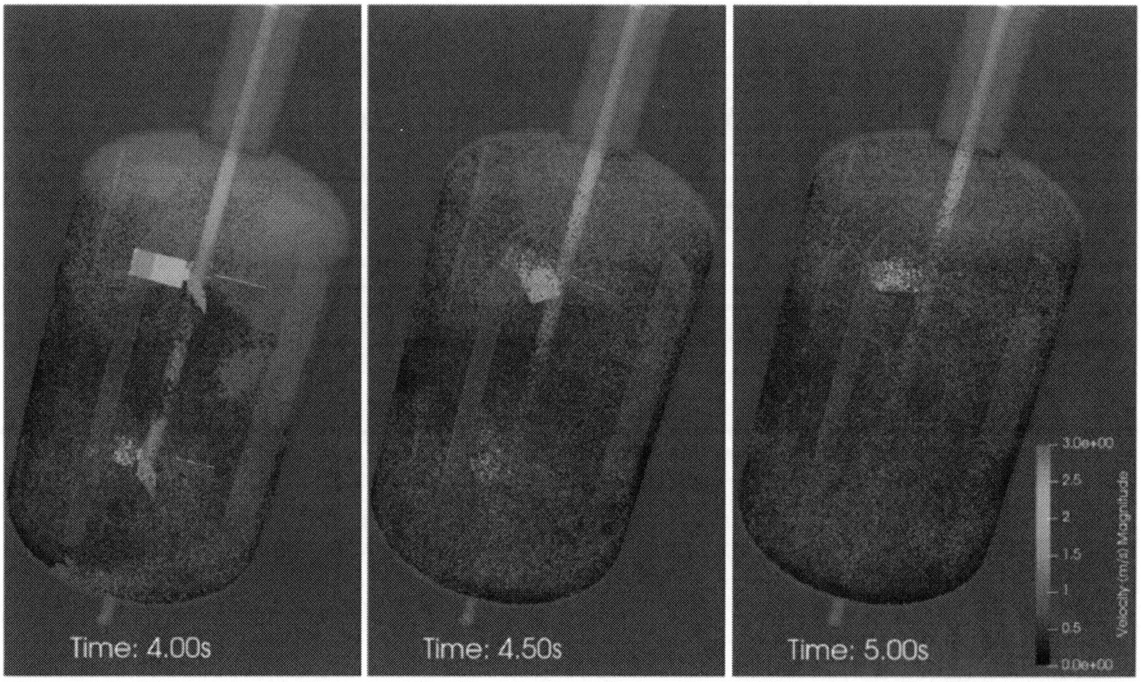

Figure 4. Tracer Particles at three Simulation Times for the Commercial Stirred Vessel

Based on the results of the initial pilot plant CFD analysis, more operational details were needed to accurately model this piece of equipment. One detail that was recommended for analysis was the inherent transient nature of withdrawal from the stirred vessel. It is believed that the particles are withdrawn from the stirred vessel in pulses corresponding to downstream operational equipment. These pulses create transient exit flow which may significantly affect stirred vessel performance. The modeled transient withdrawal rate was estimated based on the withdrawal cycle frequency, keeping the cumulative flow over time the same as the previous simulation. The simulated flow rates for the initial model and the pulse exit model are shown in Figure 5. The pulse occurs periodically and lasts for roughly 12% of the total cycle. For this simulation, two complete pulse cycles were simulated. The pulse was introduced as an increased flow rate in the inlet and was tracked to the outlet.

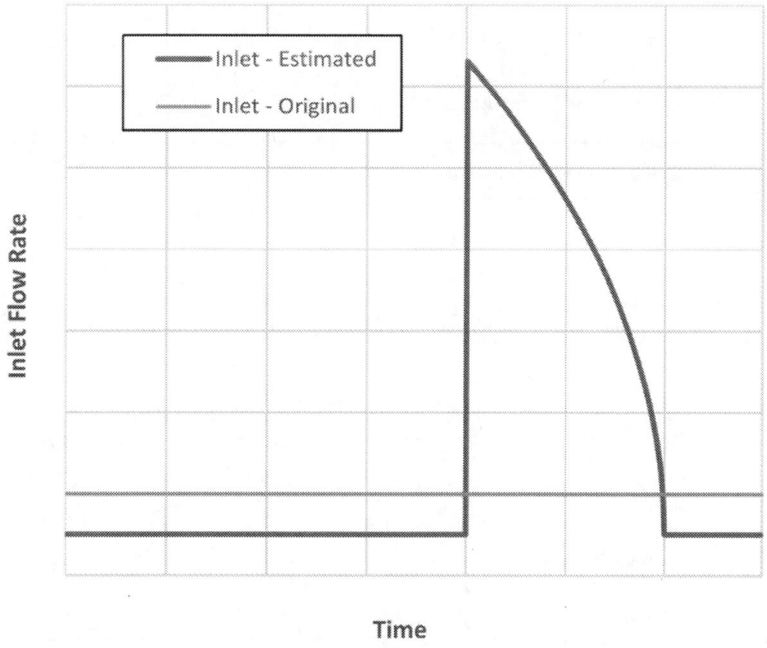

Figure 5. Inlet Velocity for the Initial and Transient Pulse Simulations

For this simulation, the iteration time step was reduced to accurately capture the transient effects of rapidly changing the velocity. The impacts of the transient pulse on the RTDs of the particles and the tracers is shown in Figure 6. As seen in this graph, the pulse changes the slope of all the particles and tracers, but there is still a significant variation based on the particle size. The CSTR for this graphic is the same one previously calculated; no effort was made to modify it based on the time-resolved inlet flow. The results of the RTDs are summarized in Table 3.

During the withdrawal, the velocities in the exit port are higher and significantly more particles of all three sizes exit the stirred vessel during this time. The exit velocity during the pulses is significantly higher than the terminal velocity of the particles. In addition, the higher flow rate also creates stronger eddies in the near vicinity of the exit in which more particles are entrained and exit the reactor. During the periods between the pulses, only a small number of the small particles exit the reactor. No significant number of medium and large particles exit the reactor between pulses.

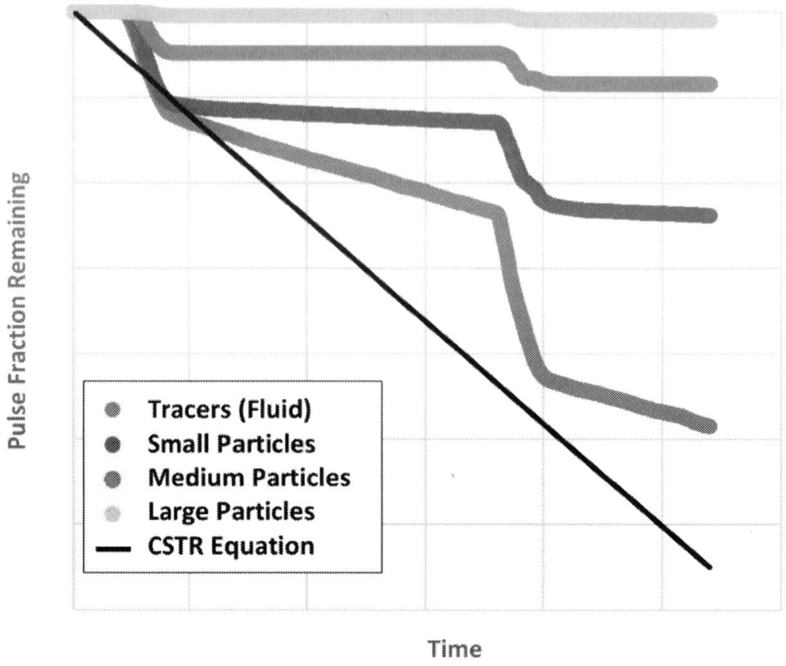

Figure 6. Pilot Plant Stirred Vessel RTDs for the Transient Pulse Simulation

Table 3. RTD Results for the Pilot Plant Stirred Vessel Pulse Simulation

Description	Initial No. of Part.	No. Exit at End Time	Comparative Mean Res. Time (min.)
Fluid (tracer)	1,000,000	9,685	**1.24**
Small	1,000,000	4,756	**3.31**
Medium	1,000,000	1,677	**10.8**
Large	1,000,000	199	**68.3**
CSTR	1,000,000	12,994	**1.00**

The main conclusion from this work is that the performance differences between the pilot-scale and commercial stirred vessel are likely the result of the particles in the pilot-scale vessel deviating from CSTR behavior, disproving the proposed hypothesis of poor mixing in the commercial stirred vessel. The next major conclusion from this work is that the particle RTDs are highly dependent on the local velocity profiles near the exit port and local turbulent eddies are important to accurately capture. The actual pulsed behavior of the pilot-scale stirred vessel is important as a CFD input to be able to accurately determine the actual RTD of the pilot plant vessel because the magnitude and duration of the pulse significantly contributes to the particle RTD. The power to perform the fully transient analysis reveals these physics that the steady state analysis does not capture.

Conclusions

A CFD study determined that the root cause of performance differences between a pilot-plant stirred vessel and a commercial stirred vessel was significant differences in the particle RTDs of the pilot-scale stirred vessel. Pilot-plant CFD results showed that the fluid RTD closely matched the CSTR RTD, but the particle RTDs varied significantly from the fluid RTD, and the particle RTDs also varied significantly with size. To model the pilot-plant vessel more accurately, a transient inlet pulse was applied. The pulsed inlet flow simulation showed significantly lower particle RTDs for the medium and large particle sizes, but only minor differences in the small particle and fluid RTDs. In contrast to the pilot-plant simulations, the commercial stirred vessel simulation showed that the fluid RTD closely matched the CSTR RTD and the particle RTDs showed much less variation and closer agreement with the CSTR RTD, with the mean residence time of the large particle size ~2x the fluid mean residence time. These results indicate that the performance differences between the pilot plant and commercial scales are likely significantly impacted by the particle RTDs. These results also indicate the need for accurate time-resolved operational data for the pilot plant, especially the outlet flow pulse, to be able to accurately determine the RTDs of the different particle sizes utilizing a CFD model.

640020 Fixed Bed Local Structure Effects on Pressure Drop and Residence Time Distribution (RTD)

Abstract:

Fixed Bed Local Structure Effects on Pressure Drop and Residence Time Distribution

Steffen Flaischlen[1,2], Jan Martin[1,2], Jule Kersebaum[1], Gregor D. Wehinger[1,2]

[1]Institute of Chemical and Electrochemical Process Engineering, Clausthal University of Technology, Clausthal-Zellerfeld/Germany

[2] Research Center Energy Storage Technologies (EST), Clausthal University of Technology, Goslar/Germany

Introduction

In slender fixed beds, the local bed structure (morphology) has a great influence on the flow properties such as pressure drop and residence time distribution (RTD) as well as heat transport [1,2]. While the particles are randomly arranged at higher reactor-to-particle diameter ratios (D/d_p>10), the structure is highly influenced by the confining wall at small D/d_p<10. The effects on the flow can be shown, for example, by pressure drop (Δp) experiments, where only a small change in morphology can lead to large changes in pressure drop [1]. Consequently, classical pressure drop correlations, where bed-averaged values are incoropated, can reproduce some of the experimental values, while for some bed arrangements, they overpredict pressure drop due to channeling. In contrast to the use of correlations, particle-resolved computational fluid dynamic (PRCFD) simulations can deal with these effects by resolving the entire packed bed structure and calculating the flow in the interstice. Therefore, new insights can be gained into structural properties such as the radial void fraction as well as the tortuosity of the packed bed and its influence on the flow fields. In addition, RTD for packed beds can be simulated. In this context, even more complex particles seem to be of interest, since with increasing complexity the mapping with simple correlations becomes more difficult.

Methods

Since PRCFD simulations require an input geometry, a geometric image of a fixed bed is needed. Synthetic generation is one of the most commonly used methods to obtain such a fixed bed representation. With the rigid body approach (RBA), it is possible to generate the fixed bed under the same filling conditions as in the experimental setup, so that the obtained structure is as realistic as possible. The whole structure is meshed using the local-caps method, in which particle contacts are cut off and filled with fluid mesh cells, which prevents the formation of deformed volume cells and thus numerical problems [4]. In addition to the averaged void fraction, local resolution is also of interest. For the analysis of this axially averaged radial void fraction, the free surface on cylindrical planes in the fixed bed is calculated. The tortuosity can be obtained by a heat transfer simulation using the heat and mass transfer analogy [5]. The residence time distribution was obtained by using a passive scalar as a massless tracer in the fluid, using a transient simulation with a concentration step at the inlet [6]. By monitoring the outlet concentration, the RTD sum curve can be calculated. On the other hand, a qualitative analysis of RTD can be performed by monitoring the tracer concentration through the bed as a video file.

Results

The investigation of different D/d_p and different particle geometries shows a strong dependence of the flow properties on the local structure. This is particularly evident in the direct comparison of two fixed beds with nearly the same reactor-to-particle diameter ratios. While for a D/d_p of 2.68 common pressure drop correlations like the Ergun or Eisfeld-Schnitzlein equations overestimate the values, PRCFD can reproduce the experiments almost perfectly. Only a small change in the geometric dimensions ($D/d_p = 2.7$) leads to an agreement between correlation and simulation, while the local structure of the bed changes extremely. The reason for this behavior is the clearly visible channel in the center of the fixed bed, which reduces the flow resistance, while at $D/d_p = 2.7$ it is closed by a particle being positioned in the center of the free space [1]. This difference can also be quantified by comparing the tortuosity of the two packed beds. The D/d_p of 2.68 has a tortuosity of τ=1.27, which means that the fluid takes a 27% longer path through the reactor compared to an empty tube (τ=1.0). In comparison, a D/d_p of 2.7 has a tortuosity of τ=1.35, showing that a small change in diameter ratio results in an 8% higher detour. In contrast to the change in tortuosity, the change in total void fraction is smaller. While the fixed bed with D/d_p of 2.68 (with the open channel) has a mean bed void fraction of ε=0.54, the bed for $D/d_p = 2.7$ results in only a 3% lower void fraction of ε=0.51. Since the pressure drop correlations depend only on the mean bed void fraction, they cannot resolve the effect of tortuosity on flow behavior. To gain additional insight into the fixed-bed structure, the RTD is analyzed. Comparing the two packed beds at the same dimensionless time q, strong differences in the tracer distribution can be seen. The fixed bed channel results in higher tracer transport in the axial direction through the center, leading to a radially non-uniform distribution of the tracer. In contrast, the fixed bed without central channel shows a more random structure and thus a uniform concentration in the radial direction. A further description can be given by the RTD sum and density curves. While the fixed bed without channeling effect can be described by a Bodenstein number Bo = 80, which is close to the value of an ideal plug flow, the fixed bed with channel shows an early curve, which is a sign for stagnant zones in the fixed bed [7].

Conclusion

It is shown that the local structure of a slender fixed bed has a great influence on the flow behavior. While the structural changes may lead to an incorrect prediction of the pressure drop, PRCFD seems to be more promising. It can therefore be used for detailed quantification of local structural effects. While the results already shown apply to packed beds of spheres, the influence of more complex particles appears to be larger due to the additional degrees of freedom in particle orientation. In addition, only the influence on flow properties has been shown so far. It can be assumed that similar effects occur in heat and mass transport. In addition, the local structures are also expected to have strong effects on reactions and thus on the conversion rate, as well as on the overall performance of the reactor.

References

[1] Flaischlen, S., Kutscherauer, M., & Wehinger, G. D. (2021). Local structure effects on pressure drop in slender fixed beds of spheres. *Chemie Ingenieur Technik*, *93*(1-2), 273-281.

[2] Dixon, A. G. (2021). Local Structure Effects on Heat Transfer in Very Low Tube-to-Particle

Diameter Ratio Fixed Beds of Spheres. *Industrial & Engineering Chemistry Research*, *60*(27), 9777-9786.

[3] Flaischlen, S., & Wehinger, G. D. (2019). Synthetic packed-bed generation for CFD simulations: blender vs. STAR-CCM+. *ChemEngineering*, *3*(2), 52.

[4] Eppinger, T., Seidler, K., & Kraume, M. (2011). DEM-CFD simulations of fixed bed reactors with small tube to particle diameter ratios. *Chemical Engineering Journal*, *166*(1), 324-331.

[5] Cooper, S. J., Eastwood, D. S., Gelb, J., Damblanc, G., Brett, D. J. L., Bradley, R. S., ... & Shearing, P. R. (2014). Image based modelling of microstructural heterogeneity in LiFePO4 electrodes for Li-ion batteries. *Journal of Power Sources*, *247*, 1033-1039.

[6] Wehinger, G. D., Kreitz, B., Nagy, A., & Turek, T. (2020). Characterization of a modular Temkin reactor with experiments and computational fluid dynamics simulations. *Chemical Engineering Journal*, *389*, 124342.

[7] Levenspiel, O. (1998). *Chemical reaction engineering*. John wiley & sons.

640140 Concentration and Temperature Uniformity during Catalytic Experiments: A CFD Analysis of the Berty Reactor

Abstract:

Concentration and Temperature Uniformity during Catalytic Experiments: A CFD Analysis of the Berty Reactor

Scott D. Anderson[1], Bjarne Kreitz[2], Prof. Dr.-Ing. Thomas Turek[1]

and Prof. Dr.-Ing. Gregor D. Wehinger[1]

[1]Institute of Chemical and Electrochemical Process Engineering,

Clausthal University of Technology, Clausthal-Zellerfeld, Germany

[2]*School of Engineering, Brown University, Providence, RI, USA*

Approximately 90 % of processes in the chemical industries rely on heterogeneous catalysis [1]. Consequentially, knowledge about the transport processes and reaction kinetics involved is of crucial importance for chemical engineers. Kinetic measurements are commonly conducted under differential conditions, allowing for easier extraction of reaction rates from end-of-pipe measurements. At integral conditions, axial profiles are recorded and detailed reactor models are necessary.

To overcome the drawbacks of the classical packed-bed reactor (PBR) continuously stirred tank reactors (CSTR) are used for catalytic measurements, as non-uniform temperature and concentration within laboratory reactors pose a challenge for derivation of kinetic models from catalytic measurements [2]. Two basic types of CSTR can be distinguished, where either the catalyst is placed in a rotating basket (Carberry reactor) or where the catalyst is fixed and a turbine circulates the gas-phase internally (Berty reactor) [3,4]. The Berty reactor has been characterized via experimental methods [5-7]. While these studies have confirmed its concentration uniformity via the residence time distribution (RTD) experimentally, little is known about the temperature distribution inside of the Berty reactor and very few studies have investigated it via Computational Fluid Dynamics (CFD) simulations [8].

The aim of this work is to bridge this gap of knowledge by investigating the temperature distribution inside a state-of-the-art Berty reactor by usage of CFD simulations to resolve flow, concentration, and temperature fields, respectively.

Therefore, a CFD model is established using *Siemens SimCenter STAR-CCM+*. Simulations of residence time behavior are performed using the *Passive Scalar* model and compared against experimental RTD data for model validation purposes. Further, global kinetics for the methanation of CO_2 over nickel-alumina catalysts are introduced into the CFD model [9]. Simulations of the strongly exothermic methanation of CO_2 are performed and the influence of operating parameters on temperature and concentration uniformity are assessed. The influence of relevant operating parameters (pressure, temperature, catalyst dilution, feed rate, stoichiometric ratio, rotation rate) is investigated via parameter variation. From an experimentalists point of view, uniform conditions in the catalyst bed that resemble the average reactor conditions are desirable. Therefore, the mean and maximum deviation of temperature and concentration in the catalyst packed bed with respect to the gas phase were computed to assess the reactors homogeneity.

Results have shown that the conversion per pass of the catalyst bed plays a determining role in the formation of temperature and concentration deviations. High conversion per pass can lead to the release of large amounts of reaction heat, causing pronounced non-isothermal temperature profiles. Furthermore, heterogeneous conditions for the concentration profile may arise, as well. It is therefore paramount for catalytic experiments to keep the conversion per pass low, which can be achieved either via dilution of the catalyst bed or the reactor recycle ratio.

The recycle ratio is a function of turbine rotation speed, operating temperature and pressure. Increasing the turbine speed leads to higher recycle ratios. Low temperature and high pressure result in high gas density. The Berty reactor turbine can transfer more energy to the gas phase at higher densities, hence increasing the recycle ratio. Consequentially, it is especially important to ensure isothermicity when conducting experiments at low pressures.

Comparing these results with findings from experimental investigations as well as the residence time simulations reveals disparities. Even at operating conditions where the RTD confirmed ideal CSTR behavior, local hot spots of up to 20 K were found, indicating that RTD experiments alone are insufficient to ensure homogeneous conditions for catalytic experiments.

This study provides the first investigation of temperature and concentration profiles inside a Berty reactor under reaction conditions. It was demonstrated how common practice of using RTD experiments for confirmation of ideal CSTR behavior can lead to erroneous conclusions and subsequently false predictions of kinetic models derived under heterogeneous reaction conditions. CFD simulations can however assist the chemical engineer with valuable insights.

References

[1] National Research Council. Catalysis Looks to the Future. Washington DC., The National Academies Press (1992).

[2] Kapteijn, F. et al. in: "Handbook of Heterogeneous Catalysis" pp 1189–1261. John Wiley & Sons, Ltd, 1997.

[3] Berty, *J. Catal. Rev. Sci. Eng. 20*, 75–96, (1979).

[4] Carberry, J. J. *Ind. Eng. Chem. 56*(11), 39-46, (1964).

[5] Haumann, M. et al. *J. Catal. 263*, 321–327, (2009).

[6] Al-Saleh, M. et al. *Chem. Eng. J. 37*(1), 35-41, (1988).

[7] Ountaksinkul, K. et al. *RSC Adv. 10*(60), 36667-36677, (2020).

[8] Ountaksinkul, K. et al. *React. Chem. Eng.* https://doi.org/10.1039/D1RE00390A (2021).

[9] Koschany, F et al. *Appl. Catal. B, 181*, 504-516, (2016).

639618 Optimal Dispatch with MPC Using Lab Scale Arduino Hardware

Abstract:

Introduction:

Co-generation and energy storage are two effective energy saving strategies for energy-intensive industrial processes like refining and chemical manufacture. Optimizing the control of energy dispatch is needed to be competitive for future energy trends where there are intermittent power sources [1]. Many processes, such as bleaching and drying of textiles and pulps, absorption chillers, enhanced oil recovery, etc.[2], have been simulated to utilize intermittent production and storage, but full-scale experimental studies are costly. This work presents a lab-scale benchmark study with the Temperature Control Lab (TCLab) to validate several optimization and simulation strategies for dispatch optimization with Model Predictive Control (MPC) with hardware-in-loop (HIL). Time series uncertainties are paired with optimizers to optimize dispatch that best utilizes the available energy during peak pricing. These results give important data for full-scale industrial dispatch optimization opportunities.

Methodology:

The TCLab validates the performance of an electrical grid dispatcher that is designed for a hybrid energy system. Test data and noise from the Arduino-based device confirm dispatcher stability and performance. Modeling and optimization of the dispatch is handled using nonlinear solvers through the Gekko Optimization Suite and Pyoptsparse [3]. Gekko solves the dynamic optimization using nonlinear programming (NLP) and differential algebraic equations (DAE). Gates et al. indicate that simultaneous methods outperform sequential methods for cogeneration optimal control [4]. Large scale nonlinear solvers IPOPT and APOPT are accessed through the two frameworks to find optimal control parameters.

A second study includes the addition of a cooling fan with the TCLab, so that both heating and cooling are available to the dispatcher. This HIL device simulates a thermal energy storage (TES) system with active charge and discharge capabilities. The test includes synthetic data as a time series to represent variable energy resources and demands.

Results and Implications:

The resulting lab-scale HIL simulation bridges the gap between simulated and physical energy storage. The results of the study demonstrate that MPC is effective for dispatch optimization, and the dispatcher validation is preparatory work for further experimental trials with a larger-scale TES system. Performance of the dispatch with MPC is tested for various system designs. Strengths and limitations of the approach are discussed with plans for scale-up testing for a lab-scale energy storage system under forecast and model uncertainty.

References

[1] Matthias Huber, Desislava Dimkova, and Thomas Hamacher. Integration of wind and solar power in europe: Assessment of flexibility requirements. *Energy,* 69:236–246, 2014.

[2] Blanco, M. and Santigosa, L.R. eds., 2016. *Advances in concentrating solar thermal research and technology*. Woodhead Publishing.

[3] Beal, Logan, et al. GEKKO Optimization Suite. *Processes* 6.8 (2018): 106.

[4] Gates, N.S., Hill, D.C., Billings, B.W., Powell, K.M. and Hedengren, J.D., Benchmarks for Grid Energy Management with Python Gekko.

638879 Towards a Framework for Resilient-Aware Design of Distributed Manufacturing Networks

Abstract:

Distributed manufacturing networks are made up of small, geographically-distributed manufacturing facilities that are located near feedstock sources and/or demand zones [1]. They are typically used in scenarios where the feedstock or demand is decentralized across a large area, such as in biorefining [2]. Having production divided amongst multiple manufacturing facilities could help minimize performance loss during disasters because surviving facilities can take on additional production capacity during recovery. Therefore, there is incentive to design distributed manufacturing networks with the capability to resist disruptions in mind.

In this study, we proposed a framework for designing a distributed manufacturing network that considers its resilience against a specific disaster scenario. A finite set of potential facility sites with varying production capacities, as well as potential connections between supply and demand nodes are considered. A mixed-integer programming (MIP) model was developed to incorporate the aforementioned alternatives, as well as a disaster scenario that could occur at one or more potential site. A penalty on the annual net profit of the network was introduced to account for the time taken for the network to recover and level of lost performance due to the disaster scenario. The MIP was then optimized to achieve maximum profit given this penalty. This framework allows for the design of resilient distributed manufacturing networks using a metric that directly reflects economic performance. The case of methanol production from refuse-derived fuels is presented to illustrate the proposed framework.

[1] Srai, J.S., et al., *Distributed manufacturing: scope, challenges and opportunities.* International Journal of Production Research, 2016. **54**(23): p. 6917-6935.

[2] Lopez-Molina, A., et al., *An Integrated Approach to the Design of Centralized and Decentralized Biorefineries with Environmental, Safety, and Economic Objectives.* Processes, 2020. **8**(12): p. 1682.

639472 Real-Time Scheduling System for the Optimal Operation of Utility Plants with Renewable Energy Assets

Abstract:

At the COP21 in Paris 2015, it was agreed globally to limit global warming to 2°C and work towards zero net carbon emissions. This was recently reinforced when countries including the US, UK, and the EU agreed to achieve carbon neutrality by 2050. These agreements have led governments to increase reporting requirements, taxes and incentives for carbon reduction, energy efficiency and carbon capture. This has significantly mobilized many industrial companies, which are major players in CO2 emissions, to take actions in a variety of forms, including studies on retrofitting processes to become compliant, investment on renewable energy resources or even exploring the possibility of changing fundamentally the basic processes that they have relied on for many decades. This has led to a massive energy transition which is ongoing.

In 2019, the National Renewable Energy Laboratory published a report [1] which suggested that oil and gas production can leverage clean power to decrease costs and maximize production while reducing environmental impact. However, this can only happen if costs and operational barriers are properly handled. A key challenge that arises in the operations of energy systems is given by the presence of multiple interacting assets [2]. Adding renewable sources with the corresponding uncertainty along with the management of energy storages introduce a much higher level of complexity.

Even though several studies have focused on understanding the economical impact of integrating renewable assets in the Oil and Gas facilities at a design stage [3,4], how to properly operate such systems in real time, to take maximum advantage of the renewable energy sources and energy storage to reduce emissions effectively is a key unanswered question that requires a holistic view of the system [5].

With this in mind, in this paper, we introduce a real-time scheduling system for the optimal energy management of process plants while accounting for renewable energy assets. This system builds on the scheduling framework outlined in [6,7] which targets fossil-fuel-based energy systems by incorporating key features, such as built-in models for wind turbines, solar panels and PV energy storages. In order to reduce the uncertainty of the energy production of such assets, frequent schedule updates are necessary. This is accomplished by integrating the scheduling system within a digitalized environment that allows seamless data connectivity across the utility plant. The resulting autonomous system is capable of gathering data and re-parametrizing the optimization models in real time by following a moving window.

We illustrate the benefits of such comprehensive system in several case studies drawn from our experience. We focus our attention on the interplay between traditional fossil-fuel-based power and steam generators such cogeneration units, steam turbines and boilers with renewable energy sources and energy storages. In particular, we outline key challenges the system is able to address when deployed under a real-time digitalized environment.

[1] Ericson, S., Engel-Cox, J., Arent, D., (2019), Approaches for Integrating Renewable Energy Technologies in Oil and Gas Operations, Technical Report, NREL/TP-6A50-72842.

[2] Serralunga, F., Ruiz J.P., Ruiz, D., Ruiz C. (2017), Integration of Decision Tools in EMS. Computer Aided Chemical Engineering 40, 2467-2472.

[3] Azadeh, M. and Elkamel, A. (2016), Integration of Renewable Energy into Oil & Gas Industries : Solar-aided Hydrogen Production.

[4] Alnifro, M., Taqvi, S.T., Ahmad, M.S., Bensaida, K., and Elkamel, A. (2017), Optimal Renewable Energy Integration into Refinery with CO2 Emissions Consideration. An Economic Feasibility Study. *IOP Conf. Ser.: Earth Environ. Sci.* 83 012018.

[5] Ruiz, J.P., Ruiz, C. (2021), Optimal Energy and Emissions Management during the Energy Transition. Decarbonization Technology Magazine.

[6] Ruiz, J. P. (2017), Decision-Making Tool for the Optimal Planning and Scheduling of Utility Assets, AICHE Spring Meeting 2017, San Antonio, TX.

[7] Ruiz, J. P., Santollani, O. (2016), Optimal Planning and Scheduling of Utility Assets, Industrial Energy Technology Conference (IETC) Proceedings.

AUTHOR INDEX